Collins

INTERNATIONAL PRIMARY SCIENCE

Student's Book 3

William Collins' dream of knowledge for all began with the publication of his first book in 1819. A self-educated mill worker, he not only enriched millions of lives, but also founded a flourishing publishing house. Today, staying true to this spirit, Collins books are packed with inspiration, innovation and practical expertise. They place you at the centre of a world of possibility and give you exactly what you need to explore it.

Collins. Freedom to teach.

Published by Collins
An imprint of HarperCollins*Publishers*
The News Building
1 London Bridge Street
London
SE1 9GF

Browse the complete Collins catalogue at
www.collins.co.uk

10 9 8 7 6 5 4

ISBN: 978-0-00-758616-5

Contributing authors: Fiona MacGregor, Karen Morrison, Tracey Baxter, Sunetra Berry, Pat Dower, Helen Harden, Pauline Hannigan, Anita Loughrey, Emily Miller, Jonathan Miller, Anne Pilling, Pete Robinson.

British Library Cataloguing in Publication Data
A Catalogue record for this publication is available from the British Library.

Commissioned by Elizabeth Catford
Project managed by Karen Williams
Design and production by Ken Vail Graphic Design

Acknowledgements
The publishers wish to thank the following for permission to reproduce photographs.
Every effort has been made to trace copyright holders and to obtain their permission for the use of copyright materials. The publishers will gladly receive any information enabling them to rectify any error or omission at the first opportunity.

(t = top, c = centre, b = bottom, r = right, l = left)

COVER: Logutenko / Shutterstock.com
p 12 tr Science Photo Library / Alamy, p 12 b Nigel Cattlin / Alamy, p 15 t Martin Shields / Alamy, p 20 tr Emily Hooton, p 58 bl Tony Bowler / Shutterstock.com, p 85 r B Christopher / Alamy.
All other photos Shutterstock.

FSC is a non-profit international organisation established to promote the responsible management of the world's forests. Products carrying the FSC label are independently certified to assure consumers that they come from forests that are managed to meet the social, economic and ecological needs of present and future generations, and other controlled sources.

MIX
Paper from
responsible sources
FSC™ C007454

Find out more about HarperCollins and the environment at **www.collins.co.uk/green**

Printed in Italy by Grafica Veneta S.p.A.

Contents

Topic 1 Plants

In this topic you will learn about the different parts of plants. You will learn that plants need water and light to grow. You will also learn how water moves in plants and that plants need healthy parts to be able to grow well. Finally, you will learn that temperature affects plants' growth.

1.1 Parts of plants

Key words
- roots
- leaves
- stem
- flowers

Plants have different parts that do different things. These parts work together to keep the plant alive and help it to grow.

The **roots** are below ground. They hold the plant in the soil. The roots also take in water from the soil.

Leaves use sunlight and water to make food for the plant to grow.

The **stem** holds the plant up and keeps the leaves off the ground. This allows the leaves to get as much sunlight as possible. The stem also joins the leaves and **flowers** to the roots.

1   Where does the water come from for the roots?

2   Trace the path of water into the leaves.

3   What do all plants need to make food?

flower

stem

leaf

root

2

Activities

1 Look at photographs A and B. Identify the roots, stem, leaves and flowers of each plant.

A

B

2 Collect leaves from many different plants. Lay them out and examine them. What is the same? What is different?

3 Are there any leaves you can eat? Talk about this and make a list of edible leaves.

I have learned

● Plants have different parts.

● Each part of the plant has a different role.

● The main plant parts are the roots, stems, leaves and flowers.

1.2 Plant roots and stems

Key words
- tap root
- fibrous root
- weed
- tree
- trunk

Plants use their roots to hold them in the soil. Roots also take in water from the soil and transport it to the leaves. The plant uses this water to make its food.

Some roots grow deep into the soil. These are called **tap roots**. Other roots spread out widely to capture as much water as possible. These roots are called **fibrous roots**. The roots of some plants become swollen, because the plant stores food in them.

1   What are the two main roles of plant roots?

2   Look at the pictures. Which is a tap root? Which is a fibrous root? Which is a root that is storing food? Explain your answers.

3   What types of roots would you find:
   - in a desert plant?
   - in a **weed** which is growing between paving stones?
   - in a mountain plant? Explain your answers.

Activities

▲ **Trees** are plants too. As a tree grows larger, its stem grows harder and thicker. This stem is called a **trunk**. Each year a new ring forms inside the trunk. We can tell how old a tree is by counting these annual (yearly) rings. ▼

1 Dig up a weed. Draw and label its roots, leaves, stem and flowers in your Workbook.

2 In your group, make a list of the root vegetables that you eat. Draw a picture of each vegetable and label your pictures.

3 Do the tree survey on this page. Write the results of your survey in your Workbook.

Tree survey

1. Is the tree growing in a shady or a sunny place?

2. Draw its shape.

3. What does the trunk look and feel like?

4. What kinds of leaves does it have?

5. Does it have flowers, fruit or thorns?

6. Can you see any roots?

7. Write down one more interesting thing about the tree.

I have learned

- Roots hold the plant in the soil.

- Roots take in water and transport it to the leaves.

- Different types of roots do different things.

1.3 Plants need water

Key words
• minerals
• absorb
• transport

Plants take in water and **minerals** from the soil through their roots. Roots **absorb** water from the soil. The water is **transported** up the stem to the other parts of the plant.

In dry soil, plants need to gather and store as much water as possible.

1 Look at pictures A and B. Name the parts of these plants.

2 Explain the way water and minerals get to these parts.

3 What kind of roots do you think each plant has? Explain your answer.

A

B

Activities

1 Plan a fair test to show that plants need water. Discuss it with your partner.

2 Use these pictures and the table to help you plan your fair test.

	Plant 1 (no water)	Plant 2 (watered once a week)	Plant 3 (watered three times a week)
Height at start			
Height, week 1			
Height, week 2			

3 Measure and record your findings in your Workbook. Explain what your results show.

I have learned

- Plants need water to grow.
- Water is taken in through the roots and transported through the stem to other parts of the plant.

1.4 **Plants need sunlight**

All living things need food to stay alive. Food gives them **energy**.

Humans and animals have to find food, but plants can make their own. They use sunlight, water and minerals from the soil and **carbon dioxide** from the air to make their food.

Without sunlight, most plants could not make their own food. If plants couldn't make food, humans and animals would have nothing to eat.

1 What is special about plants?

2 What do plants use to make food?

3 What would happen to animals and plants if there was no sunlight?

If you put one plant in a sunny place and another in the the dark what do you think will happen?

Activities

1 Using two healthy bean plants, plan a fair test to show that plants need sunlight.

2 At the end of the week, fill in the table in your Workbook. What does the experiment show?

3 Explain how you can make sure that this is a fair test.

I have learned

- Humans and animals need plants for food.
- Plants need sunlight to make food and grow.

1.5 Plants need warmth

Key words
- warmth
- temperature

You know that plants need water and sunlight to grow. They also need **warmth**. Seeds won't grow if they are in a very cold place. The **temperature** has to be warm enough for them to start growing.

Look at these pictures. The children are setting up an experiment to see how bean seeds are affected by temperature.

Step 1: Put some wet cotton wool on a saucer.

Step 2: Place a bean seed on the cotton wool.

Step 3: Cover the bean seed with some more wet cotton wool.

Step 4: Put your bean seed in a warm place, like a sunny windowsill. This is bean **A**.

Step 5: Repeat Steps 1–3, but place the second bean seed in a cold place, like a fridge. This is bean **B**.

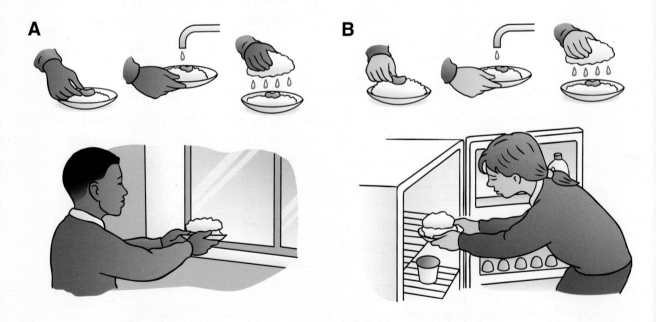

1 What is the same about what the children are doing?

2 What is different about what the children are doing?

3 What do you think will happen to bean **B**?

4 Which bean seed do you think will grow best? Why?

Did you know that scientists found some seeds in a very cold part of Canada that were 10 000 years old! They planted the seeds in a warm place and the seeds started to grow. The seeds had been waiting for 10 000 years for the warmth they needed.

Activities

1 Set up the investigations shown in the pictures on page 10. Draw the results and write your conclusions in your Workbooks on page 9.

2 Grow some seeds for lunch. Ask your teacher or your parent to give you some of these seeds: cress, lentils, chick peas or mung beans. Put the seeds in a jar of warm water for 12 hours. Follow the instructions your teacher gives you. Then enjoy fresh bean sprouts in your sandwich!

3 Imagine you are a seed that has been waiting for 10 000 years to grow. Write a paragraph explaining how you feel, as you begin to grow.

I have learned

- Plant growth is affected by temperature.
- Plants need warmth to grow.

1.6 Healthy plants

Plants need healthy roots, leaves and stems to grow well. They grow best with plenty of water and soil which is rich in minerals.

Modern farming and technology can help plants to grow well and be healthy. **Fertiliser** can be added to soil to increase the amount of minerals in it and improve the soil quality.

1 What do you think would happen if a plant had unhealthy leaves or roots? Explain your answers.

2 Look at the photographs. ▶ Why is there a difference between the plants grown with fertiliser and the plants grown without fertiliser?

3 Do all soils need fertiliser? Explain your answer.

without fertiliser

with fertiliser

with fertiliser

without fertiliser

Technology can help us grow healthy plants in areas with poor soil and a poor climate. Look at the photograph of **irrigated** fields in an area that used to be desert. ▼

4 What does the word 'irrigated' mean?

5 In what ways can we change desert land by irrigating it?

6 What problems might there be with changing desert land in this way?

I have learned

- Plants need healthy roots, leaves and stems to grow well.
- We can improve soil with fertiliser.
- We can irrigate land in dry areas.

Activities

1 Follow your teacher's instructions to make your own fertiliser.

2 Plan a fair test using your fertiliser to show the ways it improves the growth of a plant.

3 Record your findings in a table. Explain your results.

1.7 Water plants

Plants grow all over the Earth, on land and in water.
Plant parts are suited to the places in which the plants grow.

Water plants have parts that are suited to being in water.

For example, the roots of the water lily grow underwater but the leaves and flowers float on top of the water.

Look at the land plants and water plants in the picture. ▼

1 Name two plant parts that are the same in water plants and land plants.

2 In what ways are water plants the same as land plants?
   In what ways are they different?

All water plants need sunlight to make food. Some have roots that anchor them to the ground and some float around on top of the water.

Pondweed grows completely underwater. The roots grow in the mud at the bottom of the pond and the stem and leaves grow underwater. The stem grows very tall so that the leaves reach to just below the surface of the water where it is still sunny. The stem and leaves are supported by the water. Pondweed will not stand upright when it is out of the water.

Pondweed grows completely underwater. ▶

Duckweed floats on the surface of the water. Its roots do not grow in the soil. ▶

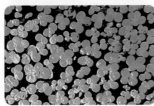

◀ *Reeds grow along the edges of a pond or river. Their roots are underwater and anchor them in the soil but their leaves and flowers grow above the water.*

3 Why will pondweed not stand upright when it is out of the water?

Activities

1 Use your Workbook to draw a section through a pond. Include four different types of water plant.

2 Complete the table in your Workbook showing the similarities and differences between a water plant and a land plant.

3 Design your own water plant. Make sure it has all the right parts. Label each part and explain what makes it suited to live in water.

I have learned

- Different plants grow on land and in water.
- Plants have parts which are suited for water or for land.

1.8 Plants in the desert

Key words
- cactus
- spine
- sap

The ghaf tree is adapted to living in very hot and dry places. It has small green leaves with a waxy covering that keeps in water. It also has very long roots to absorb as much water as possible. ▶

The **cactus** has stems that swell to hold lots of water. The folds in the stem expand when it rains, so they can store extra water. The plant stores most of this water to use when it is very hot or very dry. ▼

Because there is less water and fewer minerals in the desert, a cactus grows very slowly.

Animals in the desert also need water to survive. Instead of flat green leaves, some cactuses have **spines**. The spines stop animals eating the cactuses to get water.

▲ Aloe plants have big thick leaves that store water. The water is stored in a liquid called **sap**. Some people use aloe sap to treat scratches and sore skin.

1 Why does the cactus need to store water?

2 Where does the cactus store this water?

3 Why are the roots of the ghaf tree very long?

4 Describe one way that thorns and spines help plants to survive.

Activities

1 Make a list of desert plants that you know.

2 Make a table like the one shown below. Write how and where three different desert plants store water.

Desert plant	How and where it stores water

3 Research another plant that grows in very dry conditions. Make a presentation to your group saying what you have found out about your plant.

I have learned

● The way a plant grows is affected by temperature.

● Some plants are suited to live in very hot conditions.

● Water can be stored in swollen stems and leaves.

1.9 Mountain plants

Key words
- alpine
- needle

Some mountain environments are very cold and windy. Low temperatures mean **alpine** plants often have to live with snow or ice. Plants that grow in these conditions need to survive in the cold and wind.

Alpine plants spread out along the ground to make flat mats. They grow in thick clusters and in gaps between rocks. This gives them shelter from icy winds and lets them store as much heat from sunlight as possible.

1   What is the weather like on a high mountain?

2   Describe what the soil will be like on a high mountain.

3   Explain what makes these plants suited to live on a high mountain. ▼

▲ *These flowering plants are suited to cold, windy mountain conditions.*

The fir trees in this picture are also suited to mountainous environments. They have small **needle** shaped leaves, covered by a waxy layer to protect them from the cold conditions.

4 Why do you think it is important for alpine trees to keep warm?

5 Why do you think there are no big plants at the top of this mountain?

Activities

1 In what ways are the plants below the icy slopes suited for very cold conditions? Write one sentence about their stems and one sentence about their leaves.

2 Compare a cactus to a pine tree. What is the same? What is different? Explain your answers and write out a table in your exercise book.

3 Write a paragraph about the plants that grow in your environment. Say why they are suited to the temperature and weather in your area.

I have learned

- The way a plant grows is affected by temperature.
- Some plants are suited to live in very cold conditions.

1.10 Flowers and unusual plants

Key words
- flower
- petal
- pollen
- nutrients

Most plants have **flowers**. Flowers often have brightly coloured **petals** or have a nice smell to attract bees and other insects. When flowers are visited by bees and insects they transfer **pollen** from flower to flower.

1 Compare the flowers in the pictures. ▼ Say which parts are the same and which parts are different.

2 Where are the petals on flowers?

Some plants have very small flowers and some plants have flowers that smell very unpleasant. Some plants smell like rotting meat.

The yellow skunk cabbage flower attracts flies and beetles with its unpleasant smell. ▶

Some plants absorb **nutrients** in an unusual way. The pitcher plant has a strong smell that attracts flies. The flies fall into the pitcher plant and die. The plant uses the nutrients from the dead flies to grow and stay healthy.

3 What are nutrients?

4 Do all plants get their nutrients in the same way?

A pitcher plant. ▶

Activities

1 Draw and compare two different flowers in your Workbook.

3 Take two or three different flowers and carefully pull off the petals. Count them and draw a bar chart in your Workbook.

2 Use a press like the one in the picture to press and dry some flowers. When they are dry, arrange them on a poster and write a short description of where you found each flower.

press

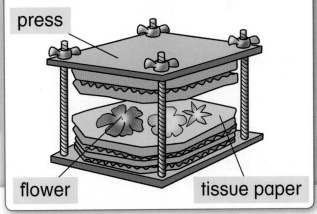

flower   tissue paper

I have learned

- Most plants have flowers.
- Flowers come in a variety of sizes, shapes and colours.

Looking back Topic 1

In this topic you have learned

- Plants have different parts.
- Each part of the plant has a different role.
- The main plant parts are the roots, stems, leaves and flowers.
- Roots hold the plant in the soil and absorb water from the soil.
- Different types of roots do different things.
- Plants need water, warmth and sunlight to grow.
- Water is transported through the stem from the roots to other parts of the plant.
- Humans and animals need plants for food.
- Plants need healthy roots, leaves and stems to grow well.
- We can improve soil with fertiliser.
- Different plants are suited to live in different places.
- The way a plant grows is affected by temperature.
- Flowers come in a variety of sizes, shapes and colours.

How well do you remember?

1 What would happen to a plant if it was never watered?

2 Are all plant roots the same? Describe two different types of roots.

3 Why is it important for plants to have healthy roots, leaves and stems?

4 Why are there so many different types of plants?

5 Describe two differences between a plant that grows in a hot desert environment and one that grows in a cold mountain environment.

Topic **2** Humans and animals

In this topic you will learn about the life processes of animals and humans. You will learn to describe the differences between living and non-living things by looking at these life processes. You will learn about the importance of a balanced and healthy diet and exercise and you will explore why some foods are bad for our health. You will also learn about human senses and explore how we use them to learn about our world. Finally, you will learn to sort living things into different groups by observing different features.

2.1 Life processes

Humans, animals and plants are living things. We can see living things and non-living things all around us.

1 Look at these pictures. Which ▶ ▼ are living things? Which are non-living things?

2 What do living things do? Make a list.

There are some things that humans, animals and plants can all do. We call these **life processes**. Plants need water and sunlight to make energy to grow. Like plants, humans and animals also need **nutrition** (food and water) as well as air to stay alive. Living things can also grow and have young (**reproduce**). All living things can also **move**.

3 In what ways do you think plants move?

Plants can move their leaves and flowers towards the sunlight. The movements are very small, so they are difficult to see.

4 Why do you think it's important that animals can move around?

Movement

Reproduction

Living things have **sensitivity** as well. This means they can respond to things they touch, smell, see or hear.

Growth

Nutrition

Sensitivity

Activities

1 Choose an animal and write a paragraph saying how you know it is alive.

2 In what ways do animals move? Make a poster to show the different ways animals move.

3 Investigate how many living things there are in your playground. Fill in your findings in the tables in your workbook.

I have learned

- Animals, plants and humans are living things.
- The life processes include movement, growth, nutrition, sensitivity and reproduction.

2.2 Living and non-living things

All living things display the life processes.
Even plants can move towards light over time. If we observe plants and animals we can see these life processes.

What about non-living things? Can we use the life processes to decide if something is living or non-living?

1 Look at the pictures. ▼
   What is living and what is non-living?
   Explain your answer using the life processes.

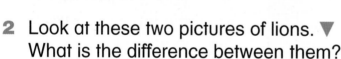

2 Look at these two pictures of lions. ▼
   What is the difference between them?

Remember, living things do all of the life processes. Non-living things might do some of the processes but will not be able to do all of them. For example, air and water can move, but they cannot reproduce.

Living and non-living things together make up the world around us. Plants need light from the sun and water from the soil to grow. Animals need air to breathe and water to drink. Animals and people need plants or other animals to eat.

Some things were once alive. A wooden gate is non-living, but the wood was once alive as it came from a living tree.

3 Is a wooden gate a living thing? Explain your answer.

Activities

1 Study the playground. Make a table like this of everything you see.

Living things	Non-living things	I'm not sure

Put anything you are not sure about in the third column.

2 How did you decide which things were living, and which were non-living? Write down the questions you asked, for example 'Does it move?'

3 Is fire alive? Use your knowledge of the life processes to explain your answer.

I have learned

- Some things are living, and some are non-living.
- Plants, animals and people are living things.
- Rocks and soil are non-living things.

2.3 Food for energy

All living things need energy to stay alive. Right now your body is using energy to read, to breathe and to grow. This energy comes from food. Without the energy your body gets from food, you would not be able to stay alive, to move or to grow.

Animals, humans and plants get food from different places. Animals and humans get food from other animals and from plants. For example, we eat meat from some animals, and we eat crops from different plants, such as potatoes or carrots. Plants get food from nutrients in the soil and from sunlight.

Different kinds of foods provide different things for the body. Look at the five main food groups – **protein**, **carbohydrates**, **fats**, **vitamins** and minerals, and **sugars**.

Carbohydrates

- Give you energy

Bread   Pasta
Rice   Potatoes

Proteins – • Help you to grow strong hair and nails

Milk   Meat   Fish
Eggs   Nuts   Cheese
Peas and Beans

1 Which foods build strong muscles and bones? To which food group do they belong?

2 What kind of foods are bread and pasta? In what ways do they help the body?

3 Can you see any foods in more than one group?

Fats

- Give you energy
- Protect internal organs
- Keeps the body warm in cooler temperatures

Butter   Olive oil
Ghee   Cheese

Vitamins and minerals

- Help to heal wounds
- Build strong bones and teeth

Fruit   Milk   Vegetables   Fish

Sugars

- Give you energy

Sweets
Fruit

Activities

1 Look at the pictures of some different foods in your Workbook. Sort them into the five different food groups.

2 In your group, look at all the pictures of meals. Which meal is the best? Draw a group picture of what you think the best meal is, and tell the class why you think it is the best.

3 Analyse the contents of a kitchen cupboard. Sort a list of the foods in the cupboard, under these headings: protein, carbohydrates, fats, vitamins and minerals, and sugars.

I have learned

- Food gives us energy and helps us to grow.
- There are different food groups.

2.4 Eating the right food

Key words
- healthy
- fibre
- balanced
- diet
- food pyramid

Your body needs a variety of foods to stay **healthy**. You need to eat carbohydrates, protein and fats. You also need vitamins, minerals, **fibre** from cereals and bread, and water.

Eating the right mix of food groups is called having a **balanced diet**. When our diet is balanced, it means that we are not having too much of one food group. It is important to eat a balanced diet because eating too much sugar or fat can damage our health.

1 Look at the picture. ▶ Do you think the girl's meal is a balanced one?

2 Which food groups can you identify in the girl's meal?

3 Explain whether or not this meal gives the girl a good variety of foods.

We can use a **food pyramid** to show the foods we should eat. The pyramid is a good way to see how much or how little of each food group we need. We should eat more of the foods at the bottom of the pyramid and less of the foods at the top.

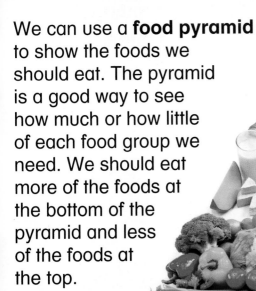

fats, oils and sweets

meat, fish, nuts and dairy products

fruit and vegetables

bread, cereal, potatoes, pasta and rice

Activities

1 Describe the food pyramid to your partner. Tell them what they need to eat more of and what they need to eat less of.

2 List the foods you ate yesterday. Draw them into a food pyramid. Did you eat healthily? What should you eat more or less of?

3 Write a healthy menu for your class for one day – breakfast, lunch and dinner.

I have learned

- My body needs a variety of foods to stay healthy.
- We should eat more of some food groups than others.

2.5 Eating the wrong food

Key words
• unhealthy
• junk food

Eating a balanced diet gives us the right amounts of energy, vitamins and minerals to move, to grow and to be strong. But eating too much or too little of one food group can make us **unhealthy**. It will make us grow too much or too little, make us sick or weak, or damage our teeth and bones.

Sugars and fats give us energy but we must be careful not to eat too much. Too much sugar can damage our teeth or make us ill. Eating too much fat can also make us unhealthy or overweight. If we do not use all the energy from sugar, it can be stored as fat in our body. Foods that have a lot of fat and sugar in them are called **junk foods**.

1 Which of the foods in the pictures are good for you?

2 Which of the foods in the pictures are 'junk foods'?

32

Eating too much fried and fatty food is also bad for you.

3 Look at this boy's meal. ▼
Say what foods you
could change to improve
his meal.

Activities

1 Cut out some pictures of food
from magazines and create
your own food pyramid poster
to display in your school.

2 List all the junk foods you
ate this week. Did you have
to eat them all? Write down
healthy foods you could
have eaten instead.

3 Scientists say we should eat balanced meals – one quarter
protein, two quarters vegetables and fruit, and one quarter
carbohydrates. Make up some good balanced meals.

I have learned

- Not all food is good for us.
- Junk food can damage our health.
- We should avoid very sweet and very fatty foods.

2.6 Exercise

Running, cycling and playing sports are all forms of **exercise**. Eating a balanced diet is very important if we want to be healthy, but a balanced diet without exercise will not make us strong or **flexible** or build up our **stamina**. When we exercise, energy that we get from food is used to build strength in our body instead of being stored as fat. Regular exercise can make us healthier, faster, stronger and more flexible.

To build your stamina, you need to exercise regularly. Running, swimming, dancing, or playing a sport like football or hockey are all good forms of exercise for building stamina. ▶

▲ *To become more flexible, you need to stretch. Can you touch your toes, like this girl?*

▲ *To increase your strength, you need to do exercises, like sit-ups and push-ups. These build up strength in our legs, arms and body.*

1 Match each word to its meaning.

strength	being bendy or supple
flexibility	being able to carry on doing something for a long time
stamina	being powerful, able to do many things

2 Give one example of a way you can improve your strength, flexibility and stamina.

3 Explain why exercise is important.

4 Look at the photos showing people doing a sit-up and a push-up. Explain the different steps to your partner.

Doing a sit-up

1

2

3

Doing a push-up

1

2

3

Activities

1 Run on the spot for 60 seconds. How do you feel? How does your skin feel? How does your heart feel? What do you feel like in half an hour?

2 Do you know some good exercises? Demonstrate them to the class. Say if they build stamina, strength or flexibility.

3 In your group design an exercise programme for Stage 3 students. The programme should allow for three sessions a week. Follow your programme for a month. Record your findings.

I have learned

- Regular exercise keeps us healthy.

- Regular exercise can increase strength, flexibility and stamina.

2.7 Your senses

Your **brain** controls your sense organs. When your eyes see something, a message is sent from them to your brain. Your brain **interprets** the message so that you can tell what the thing is. This happens very quickly so that we can react to what we see, hear, touch, taste or smell.

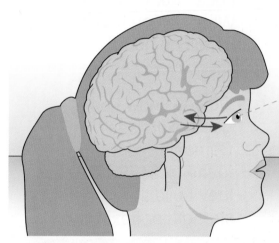

1 Why do you think it's important that our brain interprets sense information very quickly?

A balanced diet and exercise are important in keeping our sense organs healthy. Too much loud noise can damage our hearing, and reading in poor light can damage our eyesight. Our senses help us experience the world around us, so keeping them healthy is very important.

A   B   C   D

2 What message is the sense organ giving in each of these pictures? ▲

3 Which of your senses is the strongest? Explain your answer.

Your sense organs often work at the same time. When you eat a meal, you smell the food, taste the food and see the food. Your brain interprets all these signals to build an accurate image of the food that you are eating. Your brain can then remember these signals so that the next time you smell, taste or see this food you will know what it is.

Activities

1. Close your eyes and listen. Write down everything you can hear. Divide your list into sounds that are close by, and sounds that are far away.

2. Design fair experiments to test your partner's senses of taste and touch.

3. Try and communicate with your group without speaking. Is it easy for them to understand you without being able to hear you?

I have learned

- Our senses are sight, hearing, smell, touch and taste.
- Our senses give us information about the world around us.
- Our brain interprets the information for us.

2.8 How your senses help you

Your senses give you information about the world around you. This can help you to avoid danger. For example, if you see a hole in the road you can step aside, if you hear a loud noise you can cover your ears, or if you taste something bad you can stop eating it.

What happens if one of your senses is damaged, or does not work? In what ways do you have to adapt to the world around you?

Someone who cannot see is **blind**. They have to learn to read using a system called **braille**. Braille uses their sense of touch instead of sight so they can read raised 'bumps' on a page.

1 What sense are the people in the photographs missing? ▲

2 In what ways have they adapted to not having this sense?

3 What is different between the way they live and the way you live? Explain your answer.

Did you know?

Some animals have 'super senses'. Bats can find food and steer themselves at night by squeaking. The high-pitched squeak bounces off objects and the bat's brain interprets the signals like sight. Sharks can pick up very small signals given off by their prey a long distance away. All living things rely on their senses to live safely and interpret the world.

Activities

1 Imagine you are blindfolded. Make a list of what you would find most difficult on a normal day. Discuss your list in your group.

2 Visit your local Society for the Blind centre or website to find out more about ways that blind people use their other senses to make sense of the world.

3 Imagine you couldn't smell or taste anything. In what ways would this change your life? Write a paragraph to explain.

I have learned

- Our senses protect us.
- Blind people have invented many things to help them.

2.9 Classifying living things (1)

Key words
* classify
* mammal
* reptile
* insect
* antennae

We know that all living things have life processes in common. But that does not mean that all animals are the same. We can observe that a shark, an eagle and a camel are very different from each other.

1 In what ways are a shark, an eagle and a camel the same? In what ways are they different?

We can group animals together by the features that they have in common. Scientists call this grouping **classifying** animals.

2 What features can you think of that some animals have in common? Make a list.

Humans are **mammals**. Mammals have fur or hair, warm blood and give birth to live young which they feed with milk. Monkeys, elephants, horses and lions are all mammals. Another group of animals is **reptiles**. Reptiles have scales instead of fur, are cold blooded, and lay eggs instead of having live young. Turtles, crocodiles and snakes are all reptiles.

3 Look at the pictures. ▶
What groups do these animals belong to? You may have to do research. Explain your answers.

The largest group of animals on the Earth is **insects**. Insects have six legs, **antennae** on their heads and a hard outer shell covering their bodies. Insects make up more than half of all animals on the planet.

4 What features are the same about all insects?

5 What role do you think antennae have in insects?

Activities

1 Sort some animals into different groups. Name each group.

2 Choose one animal that you like. Write a paragraph about it, explaining what group it belongs to and why.

3 Do some research to find out about an insect that lives in your local environment.

I have learned

- Scientists classify animals into groups.
- Each group of animals shares common features.
- Three of the groups are mammals, insects and reptiles.

2.10 Classifying living things (2)

You have learned about mammals, reptiles and insects. There are three other main groups of animals that we can classify.

Fish have fins instead of legs, are covered in scales and live in water.

Birds have two legs, feathers, a beak and wings. They are warm blooded and lay eggs.

Amphibians have gills, like fish, but can also breathe air, often through their skins. They lay their eggs in water.

1   Look at the pictures. What do the animals have in common? ▲

2   Can you think of any other ways of classifying these animals? Explain your answers.

3   Which of these animals are fish? How do you know?

We can classify animals into even smaller groups by looking at their similar features. Monkeys and cats are all mammals, but we can see that cats can be grouped and monkeys can be grouped.

4 Why do you think scientists like to group animals?

5 Can you think of any other smaller groups within mammals?

Activities

1 Make a list of animals that you know. Group them into the main animal groups you have learned.

2 Draw some pictures to complete the table in your Workbook.

3 Look at the pictures. Research and classify these animals. Explain your answers using the features in your Workbook.

I have learned

● Other animal groups are birds, amphibians and fish.

● We can classify smaller groups within the main animal groups.

Looking back Topic 2

In this topic you have learned

- The life processes include movement, growth, nutrition, sensitivity and reproduction.
- Plants, animals and people are living things.
- Rocks and soil are non-living things.
- Food gives us energy and helps us to grow.
- We need a variety of foods to stay healthy and to have a balanced diet.
- Very sweet and very fatty foods can damage our health.
- Regular exercise keeps us healthy and builds strength, stamina and flexibility.
- Our senses are sight, hearing, smell, touch and taste.
- Our senses give us information about our world and protect us.
- Scientists classify animals into groups which share common features.
- Some main animal groups are birds, mammals, fish, reptiles, insects and amphibians.

How well do you remember?

1 What do these living things have in common?

2 How do you know that each of these animals is alive?

3 What group does each animal belong to?

4 What kind of food should animal C eat, and why?

5 What else should animal C do to keep healthy?

6 What five senses does animal C have?

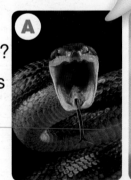
A

B

C

Topic (3) Material properties

In this topic you will discover that materials have different properties and that they can be sorted into groups based on their properties. You will learn that different materials are chosen for different purposes because of the different properties that they have. Finally, you will learn about magnetic materials and you will test materials to see if they are magnetic or not.

3.1 Properties of materials

All materials have particular properties. For example, a material can be hard, soft or shiny. Scientists study the properties of materials through experiments and observation. Colour, shape, size and **texture** are some examples of properties that scientists study. Scientists use their senses to examine the properties of objects.

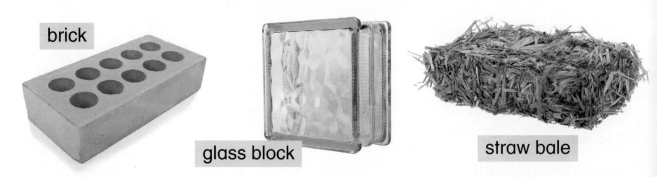

brick

glass block

straw bale

▲ Look at the picture of the brick. We can use our senses to answer questions about the different properties of the brick.

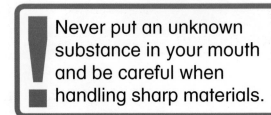

Never put an unknown substance in your mouth and be careful when handling sharp materials.

 What does it look like?
It is brown and rectangular.

 How big or small is it?
It is as long as my foot and as wide as my hand.

 What does it feel like?
It is hard, rough and heavy.

1 Now think about the properties of the other two objects next to the brick. Answer these questions for each of the objects.

   a What does it look like?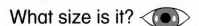

   b What size is it?

   c What does it feel like?

Here is a table of some properties that some objects can have and their opposites. You can use a table like this to help you describe the properties of different objects. You can also use this list to help you group objects.

Property	Opposite
hard	soft
big	small
smooth	rough
strong	weak
flexible	rigid
heavy	light
absorbent	waterproof
see-through	not see-through

Activities

1 Study the photographs on this page. What are the main differences between the objects in the photographs? Discuss this in your groups.

2 Choose two objects and write a description of their properties, using the questions and the properties table in this unit. Record the information in your Workbook.

3 Describe an object by its properties and see if your group can guess what it is.

I have learned

- Materials have particular properties.
- Different materials have different properties.

3.2 Hard or soft?

Scientists can group materials by looking at their properties. Materials with similar properties can be grouped together.

Look at these pictures. Some of the materials in the pictures are soft and some are hard. ▼

1 Which materials are harder than the others?

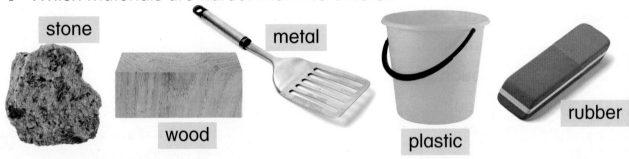

stone

wood

metal

plastic

rubber

2 Name some other hard materials. Name some other soft materials.

Look at the **surfaces** in these pictures. ◀▲

3 Which surface is the hardest? Why does this surface need to be harder than the others?

4 Describe a way you could test which surface is the hardest. What can you do to make sure the test is fair?

A hard material can scratch a soft one. The hard material is unchanged, but the soft material is scratched. You can use the rub test or the **dent** test to test the hardness of a substance.

Doing the rub test

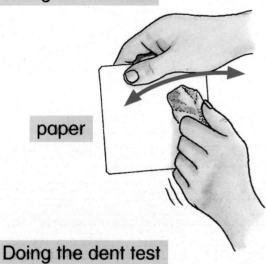

paper

Doing the dent test

stone

Activities

1. Why is it useful to use hard materials or soft materials? Write down two examples of substances that need to be hard and say why. Do the same for two substances that need to be soft.

2. Do the rub test and the dent test on the substances your teacher gives you. What could you do to make sure this is a fair test?

3. Investigate surfaces in your classroom. Find another way to test the hardness or softness of a surface without damaging it.

I have learned

- We can group materials by their properties.
- Hardness and softness are properties of materials.
- Some materials are harder than others.
- Some surfaces are harder than others.

3.3 Strength

Strong materials do not break easily when **loaded** or stretched. This means that they can carry heavy weights or support large objects. A material does not have to be hard to be strong. A plastic shopping bag, for example, is quite strong, but it is not hard.

This man has chosen the wrong shopping bag. Can you help him?

1 Why did the man's bag break?

2 Which bag is the best bag to carry shopping in? Why do you think it is the best?

3 Look at the pictures. Which materials do you think are strong and which are hard? Are there any materials which are both? ▼ ▶

50

4 Why do you think it is useful for materials to be strong but not hard? Give an example to explain your answer.

Activities

1 Set up this experiment to test whether natural or made materials are stronger. What will you do to make this test fair?

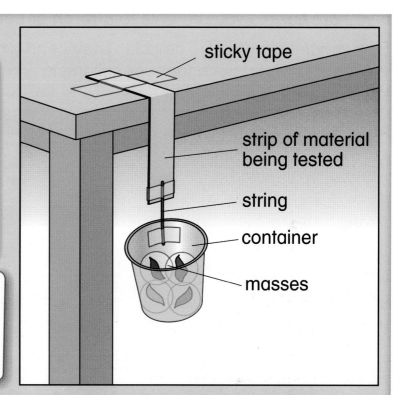

sticky tape

strip of material being tested

string

container

masses

2 Record your results in the table on page 44 of your Workbook.

3 Write a list of all the materials in your classroom. Rank them in order from the weakest to the strongest. Share your findings with your partner. How many materials are ranked the same?

I have learned

- Strong materials do not break easily when loaded or stretched.
- Strength is not the same as hardness.

3.4 Flexibility

In Topic 2 you learned about keeping your body flexible. If you are flexible you are able to **stretch** or bend easily.

Some materials are also flexible. They bend when you squash them, push them or pull them. When you stop, they spring back into shape. Scientists call this property flexibility. If something can't bend at all, they say it is **rigid**.

Look at this boy. ▶

1 What is he doing?

2 What happens to balloons when you let air out?

3 Do all stretchy materials go back to the same shape, when you stop stretching them? Explain your answer.

Many materials are chosen for a particular job because they are flexible. Clothes need to move with your body's movements, so they are made of flexible material. You need to be able to turn the pages of this book, so it is made of flexible paper.

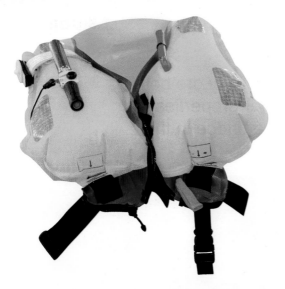

Activities

1 Identify one object in the classroom that needs to be made of a flexible material and one object that needs to be made of rigid material. Explain to your group what they are made of and why.

2 Rank some objects according to how flexible they are. Record your results in your Workbook.

3 Do an investigation to test different materials to see how flexible they are. Record your results in a table.

4 What important properties does an inflatable life jacket need to have?

5 Name a material that would not be suitable for an inflatable life jacket.

6 Inflatable life jackets are made from special nylon. Is this the most suitable material for the job? Why?

I have learned

● Flexible materials can bend, stretch or be squashed.

● Rigid materials cannot bend or change shape.

3.5 Structures

Wood and metal share some properties. They are both hard and they can both be rigid. But other properties of wood and metal are different. Why is the bridge in the photograph made of metal, rather than wood?

1 Name some main uses of wood and metal.

2 Describe some properties of wood and metal.

3 Why are wood and metal chosen for particular jobs? Give examples to explain your answer.

All buildings need to be strong and **stable**. It's important that buildings do not **collapse**.

Sometimes the shape of a building can help to make it more stable.

Very tall buildings are called skyscrapers. These are much taller than normal houses. How do you think this affects what materials are chosen to build them?

Look at the picture. ▼

4 In what ways are the buildings in the picture the same? In what ways are they different?

5 Identify one building that is more stable than others. Why do you think it is more stable?

6 What materials have been used to make these buildings? Why have the builders chosen these materials?

Activities

1 List all the things made from wood and from metal in your classroom. Discuss what properties each material has. Decide why wood or metal was used to make that object.

2 Which shapes are more stable than others? Design a fair test to find out.

3 Research the materials and shapes that builders use in hurricane areas.

I have learned

- Wood and metal are used in buildings.
- The shape of a building can make it more stable.

55

3.6 Uses of materials

Materials are used for different **purposes**. Some materials, such as cotton or nylon, are used for clothing because they are light and flexible. Other materials, like plastic, are used for storing food and drink because they are light and flexible.

Many different materials are used for **constructing** buildings because they have useful properties.

Look at the picture of the house. ▲

1 What material do you think the walls of the house are made of? Why do you think this material was chosen?

2 What material do you think the windows are made of? Why do you think this material was chosen?

3 What material do you think the door is made of? Why do you think this material was chosen?

All materials have more than one property. Glass can be smooth and see-through; metals can be hard and shiny; plastic can be light and flexible. When people choose a material to make something, they have to think about all of these properties.

4 Which material would you use to build a roof – metal or stone? Explain your answer.

5 Which material would you use to build a windscreen – plastic or glass? Explain your answer.

Activities

1 Make a list of some materials that are used to make a car. Next to each material explain why it has been chosen for that part.

2 Design a 'house of the future'. You can use any materials you want, but you must say what properties they have that make them suitable for the job.

3 Invent a new material that has more than one useful property. Describe what it could be used for.

I have learned

- Materials are chosen for particular purposes, based on their properties.
- Materials can have more than one property.

3.7 Staying the same shape

Key words
- original
- elastic

Some materials stretch more than others. Materials that stretch and then go back to their **original** shape are called **elastic** materials. Elastic materials are used in many items that we use every day.

Look at the picture of the boy. ▶

1 Why are some clothes made with elastic materials?

2 Why do you think it is useful for materials to return to their original shape after stretching?

3 Give some examples of different objects that have elastic materials in them.

Look at these pictures. ▼

4 Why do you think elastic materials are useful in many sports?

Not all elastic materials stretch as much as elastic fabric does. Some materials and objects are only slightly elastic, like a plank of wood that springs back after you stand on it. Other objects, like a squash ball, are very elastic. Tennis balls, basketballs and tennis racket strings are made out of elastic materials because they need to bounce more.

A

B

C

D

E

F

G

Activities

1 Talk about the objects in the pictures. Rank them from most elastic to least elastic. Discuss why the material being elastic is useful for each object.

2 Test a collection of materials in your group. How far does each material stretch? What do you need to keep the same to make the test fair? Record your findings in your Workbook.

3 Research an elastic material that interests you – it could be Lycra, or a balloon, or silly putty. Make notes, and prepare to tell the class about your interesting material.

I have learned

- Elastic materials can stretch and then go back to their original shape.
- Materials with elastic properties are very useful.

59

3.8 Floating or sinking?

Some objects **float** and some **sink**. The material
we use to make an object is important if we want to
make sure that the object will float. If we want to build a canoe,
a yacht or a large ship, we need to know what materials to use.

The shape and weight of an object can also make a difference to
whether it floats or not. Some objects that do not float when they
are one shape, might float when they are another shape.

1 What properties of a
   material help it to float?

2 What factors, apart
   from materials, can
   affect how well a boat
   will float?

3 Explain how big metal
   ships, like the ones in
   the picture, can float.

Activities

1 Look at the pictures of the objects at the bottom of these pages. Do you think they will float or sink? Write down your predictions in a table like this, and then test each object. Was your prediction correct?

Object	Prediction	Did it float or sink?

2 Build the paper boats in this picture. Then float them all in shallow water. Which shape of boat floats the best? Explain why.

What can you do to make a boat that floats better? Write down your plan. Then build the boat. Was your plan successful? Say why or why not.

3 Design an oil tanker. Draw the shape, and describe the materials used. Explain how your oil tanker floats.

I have learned

- Some objects float and some sink.
- Shape can help an object to float.

61

3.9 See-through or not?

Some materials are useful because of their strength or flexibility. Sometimes, materials are useful because of the way they look. **See-through** materials let light through them, so we can see objects on the other side.

Look at picture. ▶

1 Identify the things that are see-through.

2 Is all glass see-through?

3 Describe some other uses of see-through materials.

See-through glass is very important for windscreens on cars. Drivers need to see clearly, so metal or wood windscreens would not work. Plastic can be see-through. Why do you think windscreens are made of glass and not plastic?

▲ *See-through*
Lets all light through.

▲ *Cloudy*
Lets some light through.

▲ *Not see-through*
Blocks all light.

In some places we only want to let a little light in. **Cloudy** glass is used in bathroom windows, so that no one can see in. Coloured glass lets a little light in too, and can be used for decoration or drinks bottles.

4 Think of some other ways that see-through and cloudy materials are used.

Activities

1 Sort these into see-through, cloudy and not see-through bottles. Say why you think each bottle has to be see-through, cloudy or not see-through. (Look at the role of each bottle.)

2 Why are plastic bottles often used instead of glass bottles? Which material is more useful for making bottles – plastic or glass? Explain your answer.

3 What kind of material would make a good window blind? Test several materials.

I have learned

- See-through materials let light through.
- Cloudy materials let some light through.
- Not see-through materials do not let any light through.

3.10 Wet or dry?

Materials that **soak** up water are called **absorbent** materials. Materials that absorb water can be very useful if we want to soak up spills or store water for later. Some natural materials like soil can be very absorbent.

1 Why do you think absorbent soil might be useful?

Look at the pictures. ▼

sand

stone

glass

paper

sponge

fabric

wood

2 Which of these materials do you think are the most absorbent?

3 Which of these materials do you think are the least absorbent?

4 Have you used any of these materials to soak up water?

Absorbent materials can be very useful, but can you think of an example where absorbing water might be bad? Life jackets that absorbed water would become heavy and dangerous. Materials that do not let water in are called **waterproof** materials.

Look at the pictures. ▼

5 Why is it useful for each of these items to be made from a waterproof material?

6 What do you think would happen if each of these items was made from an absorbent material?

Activities

1 Name three absorbent and three waterproof materials. Explain why it is useful for these materials to be absorbent or waterproof.

2 Plan an experiment to test whether a material is absorbent or waterproof. What will you need to measure? What will you do to make the test fair?

3 Carry out your fair test on three materials. Make predictions beforehand and then record your results.

I have learned

● Some materials can absorb water and some can keep it out.

● Absorbent and waterproof materials can be useful for different things.

3.11 Magnets

Key words
- magnet
- attract
- magnetic

A **magnet** is a piece of metal that has a special property. It can **attract** other materials to it. If something is attracted to a magnet we say it is a **magnetic** material.

Magnets come in all shapes and sizes.

Materials behave in different ways when they are held near a magnet.

1 Are all materials magnetic?

Look at the picture. ▼

aluminium kitchen foil

aluminium can

copper wire

plastic duck

steel paperclip

steel nail

silver earring

gold ring

steel ruler

2 Which materials do you think are magnetic? What could you do to test this?

Magnets can be used to sort materials. Drinks cans and food cans are made from either steel or aluminium. When aluminium cans are recycled a magnet is used to remove all of the steel cans which are not needed.

3 How does a magnet help sort aluminium cans for recycling?

4 Why do you think some cans are made from aluminium and some cans are made from steel?

Activities

1 Use a magnet to test a variety of objects to see if they are magnetic or not. Record your results in your Workbook.

2 Use a magnet to find five magnetic objects and five non-magnetic objects in your classroom. Record your results in your Workbook.

3 Sort some aluminium and iron by using a magnet to separate the two metals.

I have learned

● Not all materials are magnetic.

● Magnets attract some types of metals.

● Magnets can be different sizes and shapes.

3.12 Using magnets

Magnets can be used
with other materials
to make them useful.
Look at this picture
of a fridge. ▶
There is a magnet in
the door which keeps
it closed.

1 Can you see the
  magnet that would
  keep the fridge
  door closed?
  Where do you
  think it is?

2 Explain how the
  magnet in the fridge
  door works.

The magnet that keeps the door closed is long, thin and flexible.
It is sealed within a strip of soft plastic. You cannot see the
magnet but it is strong enough to work through the plastic.
It works in the same way as the plastic coated paperclips you
have already investigated.

3 Why do you think a long, thin, flexible magnet is used in
  the fridge door?

The size, shape and strength of a magnet will depend on what it is being used for. Look at these everyday objects. ▶ They all use different types of magnets to make them work.

Very powerful magnets can lift heavy metal objects. Very strong magnets are used in computer hard drives and CD/DVD drives.

◀ *Computer hard drives are found inside the plastic casing of the computer. They store the information that is on the computer.*

Activities

1 Design a purse which uses a magnet to keep it closed. Draw your design in your Workbook and label all of the materials you will use.

2 Find out how the magnet in a tablet case can turn the device on and off.

3 Do some research to find out how computer hard drives use magnets.

I have learned

- Magnets can be thin and flexible, or large and rigid.
- The size, shape and strength of a magnet will depend on what it is being used for.
- Magnets are used in many everyday objects – even if we cannot see them.

69

Looking back Topic 3

In this topic you have learned

- Different materials have different properties.
- Hardness and softness are properties of materials.
- Other properties of materials include being flexible, rigid, elastic, see-through, not see-through or cloudy.
- We can group materials by their properties.
- Materials are chosen for particular purposes, based on their properties.
- The shape of a building can make it more stable.
- Some objects float and some sink.
- Some materials can absorb water and some can keep it out.
- Magnets attract some types of metals, which are called magnetic materials.
- Magnets can be different shapes and sizes.

How well do you remember?

Look at the three pictures. ▶

1 Write down what the objects have in common.

2 Make a table showing what materials have been used to make these objects, and why.

3 In what ways could these objects have been improved? Give one suggestion for each object.

Topic 4 Forces and motion

In this topic you will learn about forces. You will learn that pushes and pulls are forces and that we can measure them using force meters. You will learn that forces can make objects start or stop moving. You will learn that forces can also change the shapes of objects and explore how. Finally, you will learn about how forces, such as friction, can make objects move faster or slower or can change their direction.

4.1 Pushes and pulls

Look at the picture of the door. ▶
It will not move unless someone
pushes or **pulls** it. Pushes and
pulls are examples of forces.

push

pull

Some objects only work by
being pushed or pulled.

Look at toys in these pictures. ▼

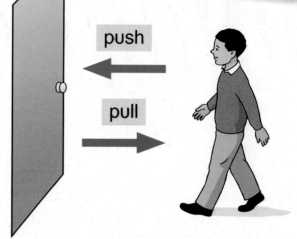

1 Divide the toys into two groups –
those you push and those you pull.

2 Which part of your body pushes or pulls the toys?

3 What happens when you push or pull the toys?

72

You can't see a force, but you can feel the result, because the object that the force acts on will move. We use pushing and pulling forces every day to do many tasks.

Look at the examples in the pictures. ▼

4 What happens when you push or pull?

5 Which do you think needs the most force?

6 What other examples of pushes or pulls can you think of?

Activities

1 Sort another group of objects that your teacher gives you into 'push' and 'pull'.

2 Look at the picture in your Workbook. Circle all of the push forces in red and all the pull forces in blue.

3 Make a list of all the push and pull forces you use on a normal school day.

I have learned

● Pushing and pulling are both forces.

● You can't see a force, but you can see the resulting effect.

4.2 Making things move

An object that is not moving is at rest, or **stationary**. What do you have to do to move a stationary object? As well as pushing or pulling objects, we can also **twist** them. Twisting objects can make them spin.

Look at these toys. ▼ ▶
What would you do to get them to move?

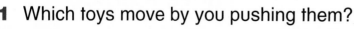

1 Which toys move by you pushing them?

2 Which toy needs a twist?

3 Which toy does not need you to touch it to make it move?

If you give something a little push, it will move forward slowly.
If you give something a big push, it will move faster.

4 What do you think will happen to each object if no force acts on it?

Look at the pictures. ▼

5 Which car is the winner?

6 Which car has moved the furthest?

7 Which car was pushed hardest?
Why do you think this?

Activities

1 Draw three objects that need a push, pull or twist to make them work in your Workbook. Then answer the questions.

2 Experiment with car races. Try using different amounts of force. Measure how far the cars have moved. Record your results in your Workbook.

3 Try the same experiment again, this time by rolling balls. What can you do to make this a fair test?

I have learned

● To move a stationary object, we need to apply a force.

● A force can be a push, a pull or a twist.

4.3 Natural forces

Objects can also move without humans pushing, pulling or twisting them. If you look out the window you might see trees moving. Wind is moving air. You can't see the wind, but you can see the way it makes trees move and you can feel it. Wind can blow sand or knock objects over. Wind is an example of a natural force. It can be a weak or a strong force.

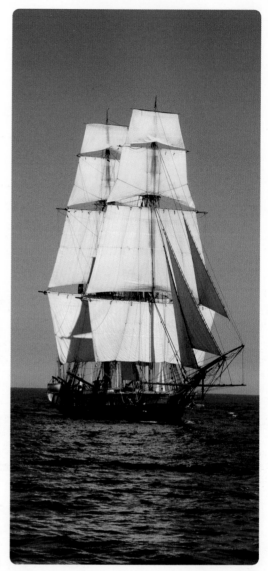

1 Which needs the strongest pushing force – the sailing ship, the boat or the sand yacht?

2 What happens if the wind stops blowing?

3 What would make the pushing force stronger?

Water can also push with a very strong force. Have you ever seen the damage a flood can do? Waves and floods can knock down even large and heavy objects.

Wind and water can also be useful forces. People have used the force of wind and water to help them for a long time. Windmills have been used to grind flour for almost 2000 years.

4 What makes the windmill turn?

5 What would make the windmill turn faster?

6 Why do you think the windmill is a very useful machine?

7 Can you think of a problem with windmills?

8 What makes a waterwheel turn?

Activities

1 Make a boat and test it to see the effect of the natural force of water. Make some waves to see what effect this has on your boat. Record your results in your Workbook.

2 Use your boat to test how the force of wind works. Record your results in your Workbook.

3 If you drop something it falls down. What makes that happen? Do some research and explain your findings to the class.

I have learned

- Wind is moving air.
- Wind and water are both forces that push.
- Wind and water forces can be used by people.

4.4 Measuring forces

Key words
- gravity
- force meter
- newtons
- newton meter
- weight

If you drop something it falls down. Why do you think this is? Remember that an object only moves if a force acts on it. What force do you think acts on an object to make it fall?

When you drop an object the force of **gravity** pulls the object down to the ground. You can measure the pull of gravity using a **force meter**.

Inside the force meter is a spring that stretches. The end of the spring has a marker that moves down the scale. This shows how strong the force is. The force is measured in **newtons**.

1 What happens to the spring when you add an object?

2 What happens to the spring when you add a heavier object?

3 What will happen if a very large object is added?

4 Why is a force meter sometimes called a **newton meter**?

The downward force of gravity acts on all objects on the Earth – people, plants, animals, all the objects in your classroom. This downward pull of gravity is called **weight**.

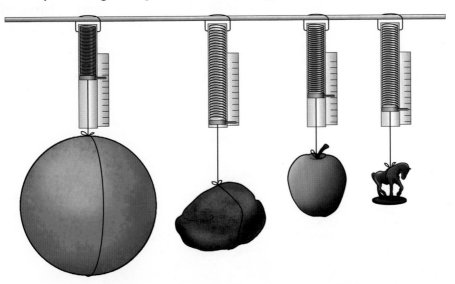

5 Which weighs more, the apple or the stone?

6 Which weighs more, the apple or the small metal horse?

7 Which weighs more, the foam ball or the small metal horse?

Activities

1 Investigate the way the length of the spring in a force meter changes as more masses are added.

2 Use a force meter to find out what different objects weigh. Record your results.

3 Imagine there was no force of gravity. Write two paragraphs describing what you think would happen without gravity.

I have learned

- Gravity is the downward force that acts on all objects.

- We can measure this force on a force meter.

- A force meter is also called a newton meter.

- Newtons are a measurement of force.

4.5 Stopping and starting

To start an object moving, a force needs to act on it.
But a force can also stop something that is already moving. Think about what happens when you push a toy car. The push force will make the car move but if the car hits a wall it will stop moving. This is because a force (from the wall) has acted on the car in the opposite **direction**.

push force

push force from wall

Fast-moving objects need a large force to make them stop. Look at the pictures of the goalkeeper catching the ball. ▶

1 In which picture was the force of the football bigger? Why?

2 In which picture does the player need to push with the biggest force?

3 Why could a ball moving very fast be dangerous?

Cars are heavy objects that move fast. If the driver of a car wants to stop, she has to apply force to the brakes. Because cars are heavy and fast-moving, the driver needs time to bring the car to a stop. You should always be very careful when crossing the road.

Activities

1 Practise throwing and stopping balls with your partner. Use a little force, then a lot. Be careful to do this safely and not to hurt each other.

2 Investigate forces with your partner, by rolling and stopping different sized balls. Record your results in your Workbook.

3 Make a road safety poster for children crossing the road. Warn children that moving traffic is dangerous, because it is difficult for cars to stop suddenly.

I have learned

- A force can stop something from moving.
- A fast-moving object needs a large force to stop it.

4.6 Changing direction

Key word
• applying

If you play a ball game, like tennis or football, you know that **applying** a force to a moving object makes it change direction. Think about hitting a ball with your racket. Does the ball travel back in the same direction? If you kick a ball against a wall, what happens?

Look at these footballers.

1 What force acts on the ball in each picture?

2 Why does the ball change direction in picture 1?

3 What does the force make the ball do in picture 2 and picture 3?

You already know that forces can be natural, such as gravity, wind or water, and that forces can be made by people. Talk about the objects and forces in these pictures. ▼ ▶

Activities

1 Investigate what happens when two marbles hit each other. Record your results in your Workbook.

2 Design your own test using two toys to see what happens when a force causes something to change direction. Record your results in your Workbook.

4 In which direction is the object moving at the start?

5 What happens to make it change direction?

6 In which direction is the object moving afterwards?

3 Copy the pictures on this page into your exercise book and draw arrows to show the direction of the object at the start and the change of direction.

I have learned

● A force can change the direction in which an object is moving.

4.7 Changing shape

Key words
- squeeze
- dent
- bend

Push down on a football or tennis ball. Observe the ball. What do you notice? The push force has changed the shape of the ball.

Forces can change the shape of some materials. You can **squeeze** a ball in your hand and it will change shape. You can stretch an elastic band and it will change shape. You can squash clay and it will change shape.

1 What is the baker doing?

2 Which two forces describe what the baker is doing to the dough?

3 What has happened to the car? Explain your answer.

4 Do all materials change shape in this way?

5 What property does the ball have?

Forces can also **dent** and **bend** objects. This can damage the objects but it can also be useful. Forces can be used to deliberately shape materials to make objects that we need.

6 Why has the nail bent?

7 How easy is it to straighten a bent nail, or remove a dent from a car?

8 In what ways do forces help make cars?

Activities

1 Work with clay to experiment with different forces. Record your results in your Workbook.

2 Talk about all the words you know relating to forces and shape. In your groups, make a diagram to summarise your knowledge.

3 In your group, research how a panel beater mends dented cars. Record your research in your Workbooks and report your findings to the class.

I have learned

● Forces can change the shape of things.

● Forces can cause damage or help make objects.

4.8 Friction

What happens if you roll a ball along the floor?
Does it keep rolling, even if it doesn't hit another object?

If you stop pedalling your bike, you will slow down. This is because there is another force acting in the opposite direction to the way the bike is moving. This force is called **friction**. Friction makes objects slow down.

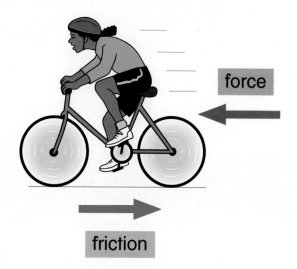

force

friction

▲ The girl is moving forward. The ground is rough. Friction is acting in the opposite direction.

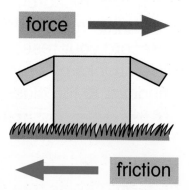

force

friction

▲ The box slides only a little across the grass. There is a large friction force acting in the opposite direction.

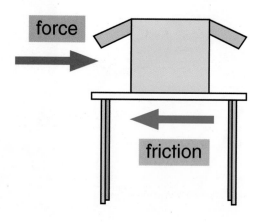

force

friction

▲ The box slides easily over the smooth table. There is a small friction force acting in the opposite direction.

1   What does 'friction' mean?

2   What would happen if the girl was cycling on an icy road?

3   Describe all of the forces in action when you push your book across your desk.

4   What happens if you push your book off the desk? Which force is acting now?

Friction can be very useful. The bottoms of trainers are designed to create friction with the ground. This helps to stop you slipping. Car tyres also have a deep tread to create friction with the road surface.

Activities

1. Copy the pictures from this page into your exercise book. For each picture draw an arrow to show the direction the object is moving and an arrow to show the friction force in the opposite direction.

2. In what ways does friction help, or not help, on the transport you use every day? Write a short paragraph to explain.

3. Research an example where friction is useful. Work with your partner and report your findings to the class.

I have learned

- Friction is a force that slows things down.
- Friction acts in the opposite direction to the movement.
- Friction can be useful.

Looking back Topic 4

In this topic you have learned

- To move a stationary object, you must apply a force.
- A force can be a push, a pull or a twist.
- Wind is a natural force that pushes.
- Gravity is the downward force that acts on all objects.
- You can measure forces using a force meter, which is also called a newton meter.
- The unit of measurement for force is the newton.
- A force can start or stop something moving, or change the direction in which it is moving.
- A force can change the shape of things.
- Friction is a force that slows things down.
- Friction acts in the opposite direction to the movement.

How well do you remember?

Look at these pictures. For each one say:
- what kind of force is being applied
- what the direction of the force is
- what the result of the action of the force is.

1

2

3

4

5

6

Glossary

absorb   To take in something, usually a liquid, or to soak it up.

absorbent   The property of materials such as sponges that will take in or soak up liquid easily.

alpine   The word used to describe the environment in high mountainous areas, and the plants and animals that live there.

amphibians   Cold blooded animals with backbones. They lay eggs and live for some of their life in water. Frogs, toads and newts are all amphibians.

antennae   The long, thin parts on the heads of insects, which are used for sensing things.

applying   Applying a force to an object means to push, pull or twist it.

attract   If something attracts objects to it, it has a force that pulls them towards it.

balanced   Containing the correct things in the correct amounts.

bend   To use force to make something curved or angular.

bird   An animal with two legs, two wings, a beak and feathers.

blind   Unable to see or sight impaired.

braille   A type of writing made with raised bumps that can be felt by blind people.

brain   The mass of nerve tissue inside your head that controls your body and enables you to think and feel.

cactus   A thick, fleshy plant that grows in deserts and is often covered in spines.

carbohydrates   Foods that contain a lot of starch or sugar.

carbon dioxide   A colourless gas that makes up part of the air.

classify   To arrange things in groups according to their shared characteristics or features.

cloudy   Not clear; you cannot see through a cloudy liquid.

collapse   To suddenly fall down or give way.

construct   To make or build something.

dent   A mark made in something by pressing it.

diet   Everything that a person or animal eats or drinks.

direction   The general line that someone or something is moving or pointing in.

elastic   The property of materials such as rubber that spring back when we stretch them.

energy   The ability to be active or of doing work.

exercise	To do physical work to improve health and fitness.
fats	Foods that are high in oily or greasy substances.
fertiliser	A substance that is added to soil to improve it.
fibre	The part of our food that we cannot digest.
fibrous roots	Thin, branching roots of a plant that spread out beneath the ground.
fish	An animal that lives in water that has a spine, gills, fins and scaly skin.
flexible	The property of materials such as rubber that you can bend easily.
float	To rest or move on the surface of a liquid without sinking.
flower	The colourful part of a plant which contains the organs from which the fruit or seeds develop.
food pyramid	A diagram that is used to represent the foods that make up a healthy diet.
force meter	A piece of scientific equipment that is used to measure forces. It measures in units called newtons. Another name for it is a newton meter.
friction	A rubbing force between two surfaces that slows things down.
gravity	The force that makes things fall when you drop them.
healthy	In good health and feeling well.
insect	A small animal with a hard outer shell, six legs and usually wings.
interpret	To work out or explain the meaning of something.
irrigate	To supply land with water brought through pipes or ditches.
junk food	Food that has little or no nutritional value.
leaves	The plural of leaf, which is the flat green part on the end of a twig or branch of a tree or other plant.
life processes	The seven life processes that are common to all living things are movement, breathing, being able to sense things, growth, having young, getting rid of waste, and taking in food.
load	To add more weight to something, for it to support or carry.
magnet	A piece of material which attracts iron or steel towards it, and which points towards north if allowed to swing freely.
magnetic	The property of materials such as iron which are attracted to a magnet.
mammal	Animals that give birth to live babies and feed their young with milk from the mother's body. Human beings, cats and whales are all mammals.
minerals	Naturally occurring substances, often from the soil, that are essential for the healthy growth of plants and animals.

move	To change place or position; to go in a particular direction.
needles	The thin, pointed leaves of some plants.
newton	A unit of force.
newton meter	Another name for a force meter.
nutrition	The process of taking in the food needed for health and growth.
nutrient	A substance that is essential for healthy growth.
original	The word used to describe the starting condition of something, often before a change of some sort.
petals	The brightly coloured parts of a flower.
pollen	A yellow powdery substance found in the flower of a plant.
protein	A food group that is made up of animal products (meat and dairy), nuts, pulses and legumes.
pull	A force that makes an object move towards you.
purpose	The reason why something is done or exists.
push	A force that makes an object move away from you.
reproduce	To produce young animals or new plants.
reptile	An animal, such as a snake or a lizard, which has scaly skin and lays eggs.
rigid	Something which is not at all flexible.
roots	The parts of a plant that grow under the ground.
sap	The juice of a plant.
see-through	The property of materials such as glass that allows us to see through them clearly.
sensitivity	The ability to respond to things or events using the senses.
sink	To fall beneath the surface of a liquid, without floating.
soak	To make something thoroughly wet.
spines	The thin prickly leaves of a cactus.
squeeze	To press something firmly from both sides at the same time.
stable	Firmly fixed or balanced and not likely to move, wobble or fall.
stamina	The ability to carry on doing something for a long time, for example physical exercise.
stationary	At rest, not moving.
stem	The long thin central part of a plant above the ground that carries the leaves and flowers.
stretch	To pull an object so it becomes longer than before.
strong	Not easily broken or damaged.
sugar	A sweet substance that plants make. It is used to sweeten food or drinks.
surface	The outside or top layer of something.

tap root	A main root that grows down vertically, for example a carrot.
temperature	How hot or cold something is.
texture	The way something feels when you touch it.
transport	To move something from one place to another.
tree	A large plant with a woody trunk.
trunk	The main woody stem of a tree.
twist	To turn one end of something in one direction while holding the other end fixed or turning it in the opposite direction.
unhealthy	Not in good health.
vitamins	Substances that are essential for health and growth.
warmth	The sensation of being warm; a moderate heat or temperature.
waterproof	The property of materials such as rubber that do not let water pass through them.
weed	A wild plant that is growing where it is not wanted.
weight	The force that is exerted on something by gravity.